What happens when
Rain Falls?

Daphne Butler

SIMON & SCHUSTER
YOUNG BOOKS

This book was conceived for
Simon & Schuster Young Books by
Globe Enterprises of Nantwich, Cheshire

Design: SPL Design
Photographs: Zefa except for
Robert Harding (7, 28)
The Image Bank (front cover, 13)

First published in Great Britain in 1993
by Simon & Schuster Young Books
Campus 400, Maylands Avenue
Hemel Hempstead, Herts HP2 7EZ

© 1993 Globe Enterprises

All rights reserved

Printed and bound in Singapore
by Kim Hup Lee Printing Co Pte Ltd

A catalogue record for this book is available
from the British Library
ISBN 0 7500 1278 1

Contents

It's raining	6-7
Where does it go?	8-9
Down to the sea	10-11
Water in the wind	12-13
Rain clouds	14-15
Sun and snow	16-17
Too much rain	18-19
Too little rain	20-21
Drinking water	22-23
Recycling water	24-25
Water power	26-27
Rain words	28-29
Index	30-31

It's raining!

Millions of raindrops are beating down onto the ground, running down window panes, and trickling into all the nooks and crannies.

How does it feel to be out in the rain? What do you wear?

Everything is wet

After the rain, the roads are streaming with water. There are puddles, and every blade of grass is glistening with water droplets.

What happens to all this water? Where does it go to?

Down to the sea

The rain soaks into the ground. It gathers into streams which tumble down the hillside, into the valleys and on down to the sea.

The sea is enormous. It covers about two thirds of the Earth.

Water in the wind

Wind blowing around the Earth picks up moisture from the sea. If the air cools, then the moisture may turn into clouds.

Not all clouds are the same. Some are thin and wispy and very high in the air. Others are white and fluffy and much closer to the ground.

Fog is cloud near the ground.

Rain clouds

Not all clouds make rain, only the ones that are dark and flat at the bottom. They usually tower high into the air above.

Inside the rain clouds, winds are rushing wildly up and down at great speed, turning the moisture back into droplets of water.

The droplets fall to the ground as rain.

Sun and snow

Sometimes when the sun shines
through the rain, you will see
a rainbow in the sky.

Each raindrop is like a tiny
prism splitting the sunlight
into a range of colours.

Sometimes, if it is very cold,
the rain falls as white fluffy
flakes of snow.

Too much rain

When it rains hard for a long time, the rivers may flood. Animals are trapped in the fields and houses may be awash with water.

In very hot countries, it rains hard each afternoon in the rainy season.

Too little rain

When no rain falls the ground dries out. Cracks appear as it shrinks. Plants die when their roots can no longer find moisture in the soil.

Some countries are always dry. The land is mostly desert. If plants are to grow at all, they must be watered each day.

The water often comes from wells deep under the ground.

Drinking water

All living things need water to drink—without water they die. People need very clean water. It must be purified so it is free of germs.

Where water is scarce, people save it in every way they can.

In the desert, it is quite chilly at night and dew forms. It is collected in tall towers.

Recycling water

Water is too precious to waste even in countries where there is plenty of rainfall.

Extra water is saved in reservoirs so there is always enough when it is needed.

Used water is called sewage. It is cleaned up on sewage farms before being returned to rivers and reservoirs.

Water power

A torrent of water rushing down
a mountain has a lot of energy.
People have learnt to use this energy.

One way is to build a dam across
a valley high in the mountains.
The water has to run out of the
dam through huge tubes.

There are wheels inside the tubes
called turbines. The water turns
the turbines and in this way
makes electricity.

Rain words

condensation The way in which water comes out of the air.

dew Water that condenses out of the air when it gets cold at nightfall.

evaporation The way in which water is taken into the air.

frost Frozen dew.

hailstones Hard frozen balls of rain.

lightning A spark of electricity from clouds during a storm.

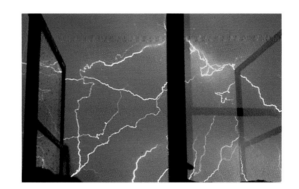

moisture Wetness.

purify To clean water so that it has no dirt and germs in it.

rainforest Huge forests in hot countries where it rains most days.

sleet Frozen rain.

Index

Aa
air 12, 15, 22
animals 18

Cc
clouds 12, 15
cold 16
colours 16
condensation 28

Dd
dam 27
desert 21, 22
dew 28
dirty water 24
drinking water 22
droplets 9, 15

Ee
Earth 11, 12
evaporation 28

Ff
flood 18
fog 12
frost 28

Gg
ground 7, 11, 12 15, 21

Hh
hailstones 28

Ll
lightning 29